CHROMIUM

A REMARKABLE MICRO-NUTRIENT WHICH MAY PROTECT AGAINST CARDIOVASCULAR DISEASE, DIABETES, AND OBESITY

Woodland Publishing Inc.
P.O. Box 160
Pleasant Grove, UT 84062

The information contained in this booklet is for educational purposes only. Please consult a professional health care physician.

TABLE OF CONTENTS

CHROMIUM: AN INTRODUCTION

Chromium deficiencies have been grossly understated and may be responsible for some of the most devastating degenerative diseases we face as a society. Because chromium is not easily absorbed and is readily excreted, learning more about chromium supplementation is critical to the maintenance of our health. The vast majority of Americans who eat poorly should be using chromium supplements to take advantage of chromium's marvelous protective and health promoting properties.

The human body contains approximately six grams of chromium, an essential micronutrient. It is primarily found in the kidneys, spleen, hair and testes and resides in the heart, pancreas, lungs and brain as well. When certain forms of chromium are present in sufficient amounts, they have been shown to protect the body against diabetes, hypoglycemia, and atherosclerosis. In addition, chromium's role in fat burning and promoting lean muscle mass is related to its ability to control food cravings and boost protein uptake.

Chromium like so many other marvelous micronutrients is beginning to emerge as a profoundly valuable mineral. Like the role of fiber, selenium, and vitamin E, it experienced a significant dietary drop when refined flours and processed foods were substituted for whole grains, raw fruits and vegetables. Interestingly, the rise of diabetes, cancer and heart disease correlate closely with the advancement of "modern" foods and provide ample evidence that Mother Nature always knows what she's doing when it comes to whole, unaltered foods.

DEFINITION:

Chromium is an essential trace mineral which is involved in the metabolism or the breaking down of glucose. It is part of what is referred to as GTF (glucose tolerance factor), a component that helps body cells receive insulin, thereby enabling it to maintain stable blood sugar levels. Chromium facilitates the proper utilization of insulin, therefore it is recommended for anyone suffering from diabetes or hypoglycemia.

GTF CHROMIUM:

When chromium is combined with glutamic acid, glycine, cysteine and niacin, it is referred to as GTF (glucose tolerance factor) chromium and is considered readily absorbable. This chromium complex is synthesized in our bodies when these ingredients work together synergistically. Because so many of us suffer from nutrient depletion, taking GTF chromium as a dietary complex is recommended.

WHAT IS CHROMIUM PICOLINATE/ POLYNICOLINATE:

Chromium picolinate is a form of the mineral which consists of a complex of chromium and picolinic acid. Picolinic acid is a natural metal chelator which is produced in the kidneys and makes chromium more bioavailable. In other worlds, this acid enhances the body's ability to absorb chromium, which can be inhibited by the presence of competitive minerals such as iron, zinc, manganese and copper.[1] Chromium picolinate was discovered by Dr. Walter Mertz and is considered by some health care experts to be the true form of GTF chromium.

THE SAFETY OF CHROMIUM:

No toxicity has been observed in oral sources of chromium at this writing. The range in which chromium is therapeutically effective is quite narrow. An excess or overdose of chromium has a tendency to inhibit rather than enhance insulin activity.[2] Inhaling chromium particles can be harmful. Always check with your physician before taking any supplement.

WATCHWORD: If you are taking any kind of medication for diabetes, high cholesterol or any other medical disorder, do not substitute chromium unless you have been advised to do so by your physician. Taking supplements for any medical condition should be cleared with your doctor.

NOTE: In discussing the benefits of chromium, be advised that GTF forms of chromium are believed to be more effective in promoting the kinds of health benefits that will subsequently be discussed. Chromium picolinate is also considered superior to isolated chromium supplements in its absorbability.

CHROMIUM: A BRIEF HISTORY

In 1929, Dr. E. Glase and G. Halpern, German scientists noticed that extracts of yeast made the action of insulin more powerful. They concluded that chromium had the ability to lower blood sugar levels by potentiating insulin.[3] Unfortunately, a great deal of time passed before subsequent studies followed. Consequently, the link between chromium and blood sugar remained relatively unknown until 1957.

In time, Dr. Walter Mertz and Paul Schwarz observed the same phenomenon and attributed it to a new dietary complex referred to as GTF (glucose tolerance factor).[4] By 1969, they had succeeded in identifying the constituents of GTF and had outlined the role it plays in enabling insulin to stimulate cell receptors, therefore allowing for sugar to more effectively leave the blood and enter the cells. At the same time, simultaneous research discovered that chromium alone was not as effective as GTF chromium (chromium that was combined with glycine, glutamic acid, cysteine and niacin).

While studies concentrated on the effect of chromium in glucose related disorders, the fact that it significantly reduced cholesterol emerged as a bonus finding. Today, expanded research has found that chromium can help control obesity, appetite and build lean muscle mass. Clearly, chromium, especially GTF chromium is a marvelous micronutrient that most of us give little thought to. Unfortunately, like selenium and magnesium, it has been stripped from so many of our modern, altered and processed foods and must be replaced in order to ensure good health.

FUNCTIONS:

Chromium activates many of the enzymes which breakdown sugar and make proteins. Glucose also known as blood sugar provides the food which fuels our cells. Insulin is the hormone which enables our cells to take up sugar by "unlocking" cell doors, allowing it to enter. Chromium boosts the ability of insulin to do this, thereby preventing our blood sugar from climbing too high, which is technically referred to as diabetes or cascading too low, also known as hypoglycemia.

It is believed that chromium helps to keep insulin molecules bound to insulin receptors located on cell walls, thereby boosting the amount of sugar that enters the cell and is subsequently burned for fuel.

Chromium also plays an important role in the process of making cholesterol, fats and protein. Scientists are continually exploring the way in which chromium operates. Apparently, it has the capability of interacting with insulin to prevent wild blood sugar fluctuations, can lower bad cholesterol while elevating the good kind, and can boost the building of lean muscle mass while decreasing the body's fat stores.

CHROMIUM AND HIGH SUGAR CONSUMPTION

Obviously, because our society routinely ingests huge amounts of refined simple carbohydrates in the form of white flour and sugar coupled with high levels of fat, the role of chromium is more crucial than ever. Ironically, it is our lack of whole grains, and our obsession with junk food which has contributed to chromium deficiencies.

In the short space of one year, the average American eats:

- 100 lbs. of refined sugar
- 300 cans of soda pop
- 200 sticks of gum
- 63 dozen doughnuts
- 50 lbs of cakes and biscuits
- 20 gallons of ice cream
- 18 lbs. of sweets and candy
- 5 lbs. of potato chips
- 7 lbs. of corn chips, popcorn and pretzels.[5]

One can only imagine the kind of stress this puts on our pancreas, not to mention the creation of excess pounds. The human body was not designed to process such enormous amounts of simple carbohydrates which make blood sugar levels surge. Complex carbohydrates, such as whole grains take longer to metabolize, therefore they help avoid blood sugar level peaks and valleys. In any event, whether we eat white bread or wheat bread, insulin will be released and what happens after its secretion is where chromium comes in.

Chromium is a vital component of GTF, which is a substance that boost the action of insulin. When we eat a carbohydrate, blood sugar levels rise and our pancreas is alerted. It secretes insulin into the bloodstream which acts to open cell doors to take the sugar our of the blood to use as food for energy.

When this process is impaired due to a malfunction, cell doors can remain closed. Consequently, the sugar continues to circulate in the blood. As harmless as sugar looks, it can act as a potent toxin in the bloodstream and wreak all kinds of havoc. This glucose intolerance can result in artery damage, injury to the retina of the eye, nerve damage, and the slow healing of wounds.

Chromium boosts the action of insulin. In other words, it

potentiates its mechanism enabling it to more effectively transport and deliver sugar from the blood to the cells. Interestingly, if you suffer from a chromium depletion, even when insulin is available, the ability to use that insulin on a cellular level is inhibited. People suffering from adult-onset diabetes suffer from this very dilemma. They may have enough insulin, but their cells have become resistant to using it.

CHROMIUM, BLOOD SUGAR LEVELS AND DIABETES

It's hard to believe that a sweet, seeming harmless substance like sugar could act as a poison to thousands of people. To anyone suffering from diabetes, elevated blood sugar levels means trouble. Diabetes occurs when insufficient levels of insulin produced in the pancreas allow sugar levels to stay high.

Diabetes is a leading cause of death and debilitation and ranks as the third leading cause of death behind heart disease and cancer. Diabetes, like heart disease is considered a western malady which is related to dietary habits (high fat, high protein, low fiber, processed foods) and lifestyle (smoking, lack of exercise and stress). Complications of diabetes include:

- nerve damage
- vision loss
- kidney disease
- heart disease
- infection

Millions of Americans suffer from diabetes and their numbers are escalating. The National Diabetes Data Group of the National Institute of Health reports that diabetes has increased over 600 percent over the course of one generation.[6] There are several theories explaining this alarming rise in diabetes and one of them points to the deplorable amount of simple carbohydrates we consume coupled with chromium deficiencies.

GTF chromium has shown some promising results in managing diabetes. It has proven that it can improve glucose tolerance in people suffering from diabetes. Several studies with brewer's yeast have shown an improvement in the ability of insulin to handle carbohydrate ingestion. The GTF variety of chromium is especially recommended for diabetics, who may not be able to convert chromium to GTF. Urine samples of insulin dependent diabetics are significantly lower in chromium than normal people. If this is the case, then diabetics have an urgent need to consume GTF chromium through diet and supplementation.

Even if you are a Type I diabetic and require insulin injections, taking GTF chromium may help to stabilize blood sugar fluctuations. Studies which have observed diabetics who needed 60 to 130 units per day of insulin were able to reduce this amount by 20 to 40 units over a one to two month period with chromium therapy.[7]

TYPE II DIABETES AND CHROMIUM

Many clinical studies have found the taking chromium supplements can help non-insulin dependent diabetics to improve their blood sugar levels. Giving GTF chromium to adult-onset diabetics in combination with a good diet and exercise program can result in a reduction of their insulin medications. Some people who go on chromium therapy can actually stop taking their insulin supplements all together.[8]

Currently, if you have Type II diabetes you are probably taking one or more of the following drugs:

- Chlorpropamide (Diabinese)
- Glipizide (Glucotrol)
- Glyburide (DiaBeta, Micronase)
- Tolazamide (Tolinase)
- Tolbutamide (Orinase)

Unfortunately, this group of drugs which are supposed to stimulate the secretion of insulin and enhance cellular response to it are not particularly effective. Forty percent of people taking these drugs fail to sufficiently control their blood glucose levels.[9] In addition, long term use of these drugs have been strongly connected to various side effects including heart attack and stroke.[10]

WARNING: If you suspect that you have Type II diabetes or a family history of the disease, chromium supplementation may be very desirable. Remember, however, not to self-diagnose and consult your physician. Not everyone who has Type II diabetes responds to chromium therapy. Never stop taking any medication without your doctor's permission.

NOTE: Poor glucose tolerance in some children has positively responded to chromium supplementation which implies that a chromium deficiency may have been responsible for the disorder in the first place.

HYPOGLYCEMIA AND CHROMIUM

Functional hypoglycemia occurs when a carbohydrate is eaten which causes blood sugar to rise rapidly signaling for increased secretion of insulin. Too much insulin causes blood sugar to take a plunge, resulting in various unpleasant symptoms. Interestingly, functional hypoglycemia can occur in people who will eventually develop Type II diabetes. In some ways, its presence should act as a warning that glucose tolerance is impaired.

SYMPTOMS OF HYPOGLYCEMIA

- nervousness
- irritability
- faintness with tremors, cold sweats
- depression
- headaches
- drowsiness
- digestive disturbances
- forgetfulness
- intense craving for sweets
- uncontrolled appetite
- excessive yawning or sighing
- insomnia
- moods which fluctuate throughout the day

Typically, if you suffer from hypoglycemia, you will feel good right after you eat and then your mood and physical status will deteriorate two to six hours after you eat.

Chromium plays a vital role in helping to prevent hypoglycemia. One study in particular has demonstrated its ability to normalize blood sugar by using 200 micrograms of chromium every day for a three month period. Eight females who suffered from hypoglycemia were placed on this supplementation. Not only did their symptoms disappear, their glucose tolerance tests improved indicating that the number of insulin receptors on their red blood cells had increased.

This same type of reaction has been observed with brewer's yeast and brings up one of the advantages of natural therapies over synthetic drugs. Chromium has the ability to prevent both high blood sugar and low blood sugar; a seemingly impossible task for one substance.

GTF chromium has proven that it can convert blood sugar into glycogen which is stored in the muscles for future use as fuel if needed. When blood sugar drops to a certain level, as it typically does in hypoglycemics, glycogen stores are released into the bloodstream bringing sugar levels back to normal.

This action can also help prevent the mood swings and depression which commonly accompany hypoglycemia. When blood sugar drops, brain chemistry is altered and one's view of life can become significantly marred.

If you suffer from hypoglycemia, you should start taking chromium in small amounts and observe your eating patterns and subsequent symptoms. Dosages can be adjusted according to your own biochemical sensitivities.

DEPRESSION AND CHROMIUM

Anyone who suffers from diabetes or hypoglycemia will wholeheartedly agree that insulin-related disorders cause mood swings. Depression has often been linked with insulin resistance, making diabetics much more susceptible to becoming depressed. Recently, scientists have discovered that insulin is carried through the blood-brain barrier and influences brain function through insulin receptors located on brain cells. These receptors contribute to the amount of norepinephrine and serotonin utilized by these cells. These two neurotransmitters determine whether we feel happy and energetic or profoundly sad and fatigued.

Anecdotal reports of improved mood in diabetics which have been taking chromium picolinate suggests that this nutrient should be further tested for the treatment and prevention of depression which results from glucose intolerance disorders.[12]

CHROMIUM AND CHOLESTEROL LEVELS

It's fascinating to learn that when one body function is impaired, it profoundly affects so many other seemingly unrelated processes. If you suffer from a glucose impairment, your body must turn to fat sources for its cellular energy. In so doing, fats are broken down for utilization and their by-products include a lipid called cholesterol. This phenomenon may explain why so many diabetics have a higher incidence of atherosclerosis and heart disease. Consider these findings:

> "In laboratory experiments, rats fed a chromium-deficient, high sugar diet showed a dramatically increased accumulation of cholesterol in the arteries. On the other hand, when rats fed a high-sugar diet were supplemented with chromium, it significantly lowered their serum cholesterol levels and resulted in less accumulation of lipids in the arteries."[13]

A study in the *American Journal of Clinical Nutrition* reported that taking 200 mcg of chromic chloride per day by a particular test group resulted in an overall lower level of cholesterol and an increase in HDL cholesterol, which is the kind we want.[14] Interestingly, using brewer's yeast to lower cholesterol was not as effective as taking chromium supplements. Apparently, yeast sources do not have enough GTF chromium to significantly change cholesterol levels for the better.

The excellent safety profile of chromium makes it an attractive lipid-lowering agent for all of us. At this point, the evidence from several clinical studies has proven that chromium is an impressive nutrient for controlling cholesterol.

CHROMIUM, HEART DISEASE AND LONGEVITY

In 1960, Dr. Henry Schroeder of the Dartmouth Trace Mineral Laboratory discovered that the rats he had supplemented with chromium not only lived abnormally long, but after they died had no cholesterol deposits in their aortas. Twenty percent of the control group of rats, who had not received chromium, showed plaque accumulation.[15] Likewise, studies of the content of chromium in drinking water showed a similar relationship in Finland, which has increased longevity in its population.

In 1994, a follow-up study on the rejuvenatory effects of chromium was conducted. In addition to re-discovering that chromium did indeed lengthen the life span of rats, scientists found that chromium promoted brain insulin activity which helped to maintain the hypothalamus in a more functionally youthful state.[16]

Conclusions drawn from these observations are that a lack of chromium can cause blood cholesterol to rise, which subsequently lodges in arteries such as the aorta and causes premature death. The effects of chromium depletion are three-fold: blood sugar rises, blood fats rise, and metabolic processes are negatively altered.

While factors other than fatty arterial deposits can cause a heart attack or stroke, our risk of having one is greatly increased when arterial plaque is present. These irregular accumulations within the arteries make them a likely place for the development of a blood clot, which can obstruct blood flow to the heart. Chromium supplementation can help prevent this scenario.

A 1980 article in the *American Journal of Clinical Nutrition* reported that chromium depletion has been strongly implicated as a contributing cause of coronary heart disease and atherosclerosis.[17] Dr. Richard A. Passwater Ph.D. in his chromium research summarized chromium's important link to heart disease in the following three observations:

1. An exceptionally marked decline in tissue chromium levels with increasing age were found in the United States. Adult levels are found to be considerably lower than those in the Far East, Middle East and Africa.

2. The tissues of Thai have more chromium than any other group, and the incidence of atherosclerosis is very low. There are few atherosclerotic complications such as heart attacks or strokes.

3. Persons dying of heart disease have virtually no chromium in their aortas, whereas those dying of accidents have aortic chromium.[18]

What is even more interesting is that other studies have found that taking chromium can actually reverse already existing atherosclerosis. The notion that inhibiting the progression of diseases like this one can be accomplished through nutritional therapy is credible, to say the least. Consider the following quote concerning laboratory tests with chromium:

> "These experiments have shown that atherosclerosis, even when established can be reversed by treatment with chromium. After a relatively short treatment period of 30 weeks with chromium, there was a substantial regression of the atherosclerotic lesion, both macroscopically (the aortas weighed less and were freer of plaques) and in terms of cholesterol content, as compared to the group that did not receive chromium. The difference between the extent of aortic involvement in the chromium treated and non-treated groups was significant by all the methods we used."[19]

With all the cholesterol hype we hear, the bottom line is that your ratio of HDL cholesterol to LDL cholesterol is what really matters. Even if your cholesterol count is acceptable and your have too much of the LDL type, you significantly increase your chances of developing plaque deposits in your arteries. It's as simple as: HDL protects the cardiovascular system and LDL threatens it.

Chromium can elevate HDL levels while it lowers the total cholesterol count. The statistics are there and they cannot be ignored. Anyone who is concerned about cholesterol or has a history of cardiovascular disease should be taking GTF chromium. It has accrued a number of impressive credentials, however, its ability to protect against heart disease has not been adequately appreciated.

In addition to its relation to arterial disease, chromium has also been found to either reverse or delay age-related changes in the body's hormonal and neural systems. The hypothalamus is directly involved and scientists believe that when it is rejuvenated, the prospect of an extended and healthy life span is enhanced.[20]

OBESITY, FOOD CRAVINGS, FITNESS AND CHROMIUM

Because glucose intolerance is frequently found in overweight individuals, the role of chromium in helping to stabilize carbohydrate cravings and boost thermogenesis (fat burning) should not be underestimated. It's no coincidence that so many people who battle the bulge are also diabetic or hypoglycemic. A 1982 study found that women who carry excess fat in their upper bodies are eight times more likely to develop diabetes than other women.[21]

As we've previously mentioned, Chromium works with insulin to efficiently metabolize carbohydrates. In addition, it plays a role in the synthesis of fatty acids. Anyone who is interested in reducing their fat percentage and boosting muscle content should be aware of chromium and its contribution to fitness. Consider the following quotes:

> "Preliminary studies of athletes by Gary W. Evans and others have indicated that chromium picolinate was effective in increasing lean body mass (muscle) and reducing the percentage of body fat."[22]

> "The increase in lean body mass of the athletes taking chromium picolinate was 44 percent greater than the increase in lean body mass of the athletes taking a placebo. The decrease in total body composition of fat was 3.5 times greater in the men given the chromium picolinate."[23]

What these studies imply is that for the burning of fat to be optimized, chromium supplementation should be considered. If you think you're getting enough chromium through your diet, think again. It's much more likely that your system is chromium deficient.

> "New research suggests that depletion of chromium in the body might impair the development of lean body mass and contribute to excess body fat."[24]

Translating this data means that by supplementing our diets with chromium, we can lower body weight and increase lean muscle mass. Interestingly, exercising can actually increase tissue levels of chromium...an added benefit to anyone who wants to stay in shape.[25]

APPETITE CONTROL

The physiological phenomenon of hunger is thought to be triggered, in part, by the activity of a certain portion of the brain which lets us know if we're feeling full and satisfied or not. Whatever message this appestat center delivers is dependent on blood sugar availability within its own cells. It only makes sense that if chromium boosts the function of insulin, which is to get sugar into the cells, it should help to suppress hunger.

More and more research is discovering that when certain neurochemicals in the brain take a dip, carbohydrate cravings can occur. Apparently carbohydrates are metabolized into not only glucose but act as precursors to other brain chemicals like serotonin as well. The role of glucose is crucial to how we feel, both mentally and physically. False hunger or abnormal cravings for sweets are intrinsically linked to brain cell chemistry. Chromium can help to keep our appestats in control, helping us avoid the kind of binge eating and snacking that puts on the pounds.

In addition, if we are cutting calories, we may need to use glycogen reserves for extra energy and chromium helps to keep those glucose reservoirs full.

A 1994 study found that chromium directly affects the insulin activity of brain cells significantly impacting the hypothalamus, which is involved in regulating appetite and burning fat. Clearly, chromium supplementation enhances the process of weight loss as it relates to both fat burning and the craving for food.

SOME ADDED FACTS ABOUT THE PICOLINATE FORM OF CHROMIUM

CHROMIUM PICOLINATE AND LEAN MUSCLE MASS

Several studies have shown the ability of chromium picolinate to increase the lean muscle mass of athletes who took the supplement compared to those taking a placebo by 44 percent.[26] Mass increased and body fat decreased. While lean body mass was on the increase, total body fat decreased at a rate 3.5 times greater in the group taking the chromium.

CHROMIUM PICOLINATE AND INSULIN MECHANICS

Because the picolinate version of chromium is readily absorbed and enhances the mechanics of using insulin properly, it, along with GTF chromium is recommended for:

- adult onset noninsulin-dependent diabetes
- hypercholesterolemia
- hypoglycemia
- eating disorders
- weight control
- sugar cravings
- muscle building booster
- Turner's Syndrome (impaired carbohydrate metabolism)

CHROMIUM PICOLINATE AND CHOLESTEROL LEVELS

The picolinate form of chromium has demonstrated its impressive ability to lower blood fats including cholesterol. Recently, tests have confirmed that taking chromium picolinate can decrease certain harmful blood lipids such as LDL cholesterol.[27] What these studies suggest is combining chromium supplementation with a healthy low fat diet to prevent the high cholesterol levels which inevitably lead to heart disease.

CHROMIUM DEPLETION:

The average American diet may be deficient in this very vital mineral. Some experts believe that the reason the incidence of diabetes, hypoglycemia and obesity is so high in this country is because chromium is lacking. If we do not get enough chromium, we jeopardize our ability to maintain normal levels of blood sugar. Because our soil and water supplies are often chromium deficient coupled with our poor eating habits, chromium deficiencies can commonly occur.

According to Sheldon Saul Hendler, M.D., Ph.D., who wrote the *Doctor's Vitamin and Mineral Encyclopedia*, chromium deficiencies are more common in the United States than previously thought. The high consumption of refined foods and chromium depleted soils have created chromium poor societies. Some scientists firmly believe that chromium deficiencies are one of the most serious nutritional problems we face today.

Typical drinking water supplies contain only trace amounts of chromium and foods high in chromium are rarely eaten consistently enough to maintain required chromium levels in the body.

Our love affair with simple rather than complex carbohydrates has only served to compound our chromium problems. Developed countries are notoriously low in chromium. Just the process of refining and fragmenting grains removes over three quarters of its natural chromium content. As is the case with so many other health related facts, this should serve as a wake up call to most of us who have abandoned whole grains for refined ones. Mother Nature knew what she was doing long before we meddled with foods that were perfectly balanced.

To make matters worse, our chromium depleted diets also promote the breakdown of GTF chromium in our bodies resulting in its increased excretion. Just eating carbohydrates raises serum chromium levels which causes it to be expelled in the urine. Consequently, we are left with meager chromium reserves and become likely candidates for diabetes, heart disease and protein metabolism disorders.

In addition to diseases like diabetes and hypoglycemia, a lack of chromium can cause a rise in the levels of certain fats in the blood, including cholesterol, which can lead to cardiovascular disease. Whenever lipid levels rise in the blood, the risk of fatty deposits lodging themselves within our arteries is increased.

Because chromium is the major mineral involved in insulin utilization, there is growing evidence that many glucose related diseases may actually be manifestations of a chromium deficiency.

These same correlation can be made with high cholesterol levels. People from several Oriental countries who have consistently low cholesterol levels have high concentrations of chromium in their tissue samples.[28]

Athletes may find themselves at risk for chromium deficiency due to their strenuous physical activity. Energy requirements of most athletes are greatly increased and rely heavily on efficient glucose metabolism to fuel the cells.

Chromium concentrations in the body are known to

decline with age.

A LACK OF CHROMIUM CAN RESULT IN:

- retarded growth rate
- glucose intolerance (possibly diseases like Turner's Syndrome, hypoglycemia and diabetes)
- atherosclerosis
- heart disease

CAN YOU TELL IF YOU'RE CHROMIUM DEFICIENT?

Having your hair analyzed for chromium content is considered a reliable way to assess your chromium status. To the contrary, blood and urine testing for chromium is considered unreliable because it does not necessarily indicate the amount of chromium which resides in the tissues. Absorbing and assimilating chromium is difficult.

CHROMIUM REQUIREMENTS

There are no RDA requirements for chromium and a lack of data makes it difficult to assess exact amounts. Remember that chromium is poorly absorbed unless it exists in the trivalent form which is found in the glucose tolerance factor.

THE CASE FOR CHROMIUM SUPPLEMENTATION

A shocking statistic which emerged from a Canadian study indicated that as many as two thirds of premenopausal women do get the chromium they need. Many of these women had daily intakes of chromium far below the level that the U.S. Food and Nutrition Board consider safe.[29] At this writing, there is no official established RDA requirement for chromium. Some studies have indicated that 200 to 290 mcg (micrograms) daily are required to maintain a desirable chromium balance.

Estimates put the average American intake of chromium at 50-100 mcg per day. Even if you believe you are eating nutritiously, you may still be at risk for low chromium levels. Richard A. Anderson, Ph.D. a biochemist at the U.S. Department of Agriculture Human Nutrition Research Center in Maryland has said,

> "Regardless of how you cut the cards, you're not getting enough chromium in your diet. Even diets designed by dietitians don't provide nearly enough chromium."

The bottom line is that many of us are either deficient in chromium or are unable to synthesize adequate amounts of GTF within our bodies. The truth of the matter is that without enough GTF, we cannot maintain optimal health and fall prey to degenerative diseases like diabetes or heart disease. As is the case with any vitamin or mineral supplement, check with your physician before you supplement your diet with chromium.

TYPES OF CHROMIUM SUPPLEMENTS

- chromium as an individual supplement (chromium chloride)
- GTF chromium chelate
- chromium picolinate or polynicolinate (can also be part of GTF chromium)
- brewer's yeast as a source of chromium

Chromium is readily available as an individual supplement or as a constituent of certain vitamin or mineral formulas. GTF chromium which consists of chromium combined with glutamic acid, glycine, cysteine and niacin is recommended for its absorbability. The picolinate variety of chromium has been praised for its ability to be assimilated in the tissues and can be part of the GTF formula. Many forms of chromium originate from brewer's yeast, therefore if you have trouble with yeast allergies or infections, you should avoid this source of chromium.

NATURAL FOOD SOURCES OF CHROMIUM

While several different kinds of foods contain chromium, much of that chromium may not be as bioactive as GTF chromium. Foods which are highest in biologically active chromium include:

- brewer's yeast
- calf's liver
- whole wheat bread

Foods which also contain significant amounts of chromium that isn't as high in GTF function include:

- beef
- wheat bran
- rye bread
- dried beans
- oysters
- potatoes
- wheat germ
- green peppers
- eggs
- chicken
- apples
- butter
- Swiss cheese
- bananas
- spinach
- cabbage
- mushrooms
- beer

NOTE: Leafy vegetables contain chromium but it is not readily absorbed. In addition, eating white bread, flour or rice does not supply the body with chromium. The chromium that was stripped in the grain milling process is not replaced when the bread is subsequently enriched.

It isn't surprising that so many of us are chromium deficient. We rarely eat the types of foods which supply chromium.

OPTIMUM DAILY ALLOWANCE OF CHROMIUM:

- 200 to 600 mcg per day, per adult

NOTE: Treating glucose related diseases, high cholesterol levels or obesity may require therapeutic dosages which can go up to 600 mcg per day. 400 mcg is the usual therapeutic dosage for diabetics and hypoglycemics.

Other complimentary nutrients to chromium include: B-complex, copper and magnesium

PRIMARY APPLICATIONS OF CHROMIUM

- **APPETITE CONTROL**
- **FOOD BINGING**
- **DIABETES**
- **HYPOGLYCEMIA**
- **HEART DISEASE**
- **HIGH CHOLESTEROL**
- **TURNER'S SYNDROME**
- **OBESITY**
- **WEIGHT CONTROL AND FITNESS**
- **ATHEROSCLEROSIS**
- **DISTURBED AMINO ACID METABOLISM**
- **SUGAR CRAVINGS**

SUMMARY

The very vital role of chromium in health maintenance and disease prevention is just beginning to emerge. The findings of early studies that chromium deficiencies are intrinsically linked to the development of heart disease and diabetes have spawned considerable new interest in the mineral. Today we know what they didn't realize then: that dietary chromium functions best only after it has become a part of the glucose tolerance factor within the body. Hopefully, additional research will continue to track chromium and how it affects our ability to metabolize carbohydrates. Because digesting and assimilating carbohydrates is so crucial to fueling our cellular systems, glucose disorders take a tremendous toll on health and cause thousands of deaths annually.

The one thing that becomes crystal clear in studying the interrelationships between blood sugar, blood fats, obesity, heart disease and diabetes is that all of our body systems are innately inter-dependent. One malfunction profoundly affects the health of seemingly unrelated body systems and proves that learning to look at ourselves as a complex integrated whole machine is vital to understanding and providing the nutrients we need to enjoy vibrant good health and longevity.

ENDNOTES

[1]Robert H. Garrison Jr. M.A., R. Ph. and Elizabeth Somer, M.A.R.D. *The Nutrition Desk Reference,* (New Canaan, Connecticut: Keats Publishing, 1990.), 127.

[2]Ibid., 62.

[3]Richard A. Passwater Ph.D. GTF *Chromium.* (New Canaan, Connecticut: Keats Publishing, 1982), 2. See also Glaser, E. and Halpern, G. 1929. *Biochemistry* Z. 207: 377-383.

[4]Ibid., 3. See also Mertz, W. and Schwarz, K., 1955. *Archives of Biochemistry and Biophysics* 58: 504-508.

[5]Michael Murray, N.D., and Joseph Pizzorno, N.D., *Encyclopedia of Natural Medicine.* (Rocklin California: Prima Publishing, 1991), 41.

[6]Passwater, 10.

[7]Ibid., 18 See also Doisy, R. J. et al., in *Trace Elements in Human Health and Disease* Vol. 11, Prard, Oberleas, eds. New York: Academic press.

[8]Shari Lieberman and Nancy Bruning. *The Real Vitamin and Mineral Book* (New York: Avery Publishing Group Inc., 1990), 166.

[9]Michael T. Murray, *Diabetes and Hypoglycemia.* (Rocklin, California: Prima Publishing, 1994), 15.

[10]Ibid., 16.

[11]Ibid., 92. See also Anderson, R.A., "Chromium glucose tolerance and diabetes," *Biological Trace Element Research* 32: 19- 24, 1992.

[12]M. F. McCarty. "Enhancing central and peripheral insulin activity as a strategy for the treatment of endogenous depression, an adjuvant role for chromium picolinate?" *Med- Hypotheses,* (Oct; 1994 43(4): 247-52.

[13]Lieberman, 166.

[14]Garrison 110.

[15]Passwater 4.

[16]M.F. McCarty, "Longevity effect of chromium

picolinate, `rejuvenation" of hypothalamic function?" *Med-Hypothesis,* (Oct; 1994, 43(4): 253-65.

[17]Passwater, 5.

[18]Ibid., 5.

[19]Ibid., 6.

[20]McCarty, "Longevity," 253-65.

[21]Passwater, 11.

[22]Lieberman, 166.

[23]G. Otte, "The Effect of Chromium Picolinate on Lean Body Mass Development in Exercising Athletes," Bemidiji State University.

[24]Garrison, 127.

[25]Murray, *Encyclopedia,* 283.

[26]Garrison, 127.

[27]Ibid.

[28]Lieberman, 166.

[29]Garrison 111.